THE VINCENT THOMAS BRIDGE

San Pedro's `Golden Gate´

Commemorative

Edition # _2723_

*A spectacular view of the Vincent Thomas Bridge
as seen from Knoll Hill.*

THE VINCENT THOMAS BRIDGE

San Pedro's `Golden Gate´

Authored and Edited by:
Arthur A. Almeida
Irene M. Almeida
Flora Twyman Baker
Kathryn Schultz
Gayle Williamson

Special Photography by:
Leon Calloway

San Pedro Bay Historical Society

Dedication

The bridge between San Pedro and Terminal Island is the State of California's complement to the great Federal breakwater built at San Pedro Point between the years 1899 and 1912. San Pedro, the southern appendage of the City of Los Angeles known as the "The Harbor," has been blessed with these two engineering masterpieces.

In 1963, the dedication, naming and opening of the Vincent Thomas Bridge were celebrated with suitable ceremony and enthusiasm. It is altogether fitting that on the 25th anniversary of the completion of this structure the story should be retold and Thomas remembered, as he did so much to bring the green span to the harbor. It is with pride that we dedicate this history to the memory of Assemblyman Vincent Thomas.

We gratefully thank Ralph and Stanley Di Meglio, Mae Sakamoto and Mrs. Vincent Thomas for their generous support. Further, we acknowledge the assistance of Lilio E. Gattoni, Everett Gordon Hager, Louis Kanaster, the Los Angeles Harbor Department, William L. Olesen and Mrs. Vincent Thomas in the preparation of this historical publication.

Book Design: Irene M. Almeida
Typesetting and Paste-Up: The San Pedro Press
Printing: Printmaster

ISBN 0-9611556-2-0

Contents

Westerly view of the San Pedro waterfront, circa 1917-18, showing the Fifth Street Landing to the immediate left of the Southern Pacific Passenger Station. The ferry departing the dock is the A.B.

Looking north, this one cent postcard shows the Southern Pacific Depot to the left of the Main Channel. The ferry launch Blanche in the foreground has just cast off and is heading toward Berth 231 in East San Pedro.

Photo taken February 2, 1927, of the incapacitated team ferry being towed by a tug.

Discovering the Need

It has been called the "bridge to nowhere" but beginning with the arrival of the first settlers into the San Pedro Bay area there was a desire and need to cross over the small body of water separating Rattlesnake (later Terminal) Island from the mainland. They approached the island from several directions: westerly from Long Beach, southwest from Wilmington and in an easterly direction from San Pedro.

Sources have stated that Captain Martin E. Lindskow was the first to offer ferry service with a row boat from San Pedro to Rattlesnake Island in the early 1870s. The boat left from the Fifth Street Landing in San Pedro whenever there were passengers, there being no regular schedule.

In 1882, Captain Mitchell Duffy arrived in San Pedro and soon established a regular cross-channel ferry service to East San Pedro on Rattlesnake Island. By 1891, the Los Angeles Terminal Railway had purchased the island renaming it Terminal. A trestle carrying tracks was built across the harbor between the mainland and what is now called Terminal Island. The railroad built its wharf directly across the channel from the depot of the Southern Pacific Railroad in San Pedro and the two stations were linked by the ferry system.

Captain Duffy began to add to his ferryboat equipment and installed the first gasoline engine on a boat in the harbor. During the 30 years he operated his service, he built and added to his fleet the Blanche, the Dora, the Ina, the Georgie, and the Elsie. They were named after his children and all held between 30 and 40 passengers. Duffy also operated the Orient which had a capacity of one hundred passengers.

On June 6th, 1899, the Southern Pacific Railroad requested permission from the City of San Pedro to establish a ferry service across the channel. The Chamber of Commerce represented by Judge William H. Savage was against the petition. The Southern Pacific Railroad offered to deed to the City of San Pedro the ferry building and the approaches at the foot of Fifth Street in order to provide a connection with East San Pedro. The Los Angeles Terminal Railroad was also willing to deed waterfront property for a landing on the island side to the County of Los Angeles. These inducements won over the opposition and on November 9, 1899, the ferryboat Ramona, owned by the Southern Pacific Railroad, was carrying passengers across the channel.

The Islander making its regular channel crossing. The San Pedro Terminal, which now houses the Los Angeles Maritime Museum, is in the foreground. A break bulk freighter is tied up at the Berth 232 finger pier at center left, and the East San Pedro ferry depot stands to its right. Both pier and depot no longer exist; Evergreen containers are now handled in this area.

Foot passengers disembarking from the Islander through the Ferry Terminal ramp at the second level of the building.

This 1963 photo illustrates the past, present and future modes of transportation connecting San Pedro to Terminal Island. The ferries Ace and Islander are in the foreground of the bridge under construction.

With the development of Brighton Beach on Terminal Island and the growth in population of both East San Pedro and the settlement called Terminal, the need for larger boats that would also carry vehicles became apparent. By 1911, the San Pedro Transportation Co., owned by Frank Garbutt and Matt Walsh, was operating under a franchise awarded by the newly formed Harbor Commission of the Port of Los Angeles. Their team ferries, which carried passengers as well as horses, wagons and automobiles, were leaving from Canal Street in Wilmington and First Street in San Pedro to Terminal Island. The fare for a passenger was five cents one way, or two and one-half cents if a 75 cent thirty ride ticket was purchased. By 1915, the Port of Los Angeles had constructed new ferry slips at Canal Street, Wilmington; Fourth Street, Terminal Island and First Street, San Pedro.

The Harbor Department charged the operator of such landings anywhere from $10 to $50 a month for their use. Passengers crossing from San Pedro to Terminal Island included summer visitors to Brighton Beach, cannery workers, residents of Terminal Island coming to San Pedro for goods and services, and school children who had completed elementary grades on the island and were now ready for junior and senior high school in San Pedro.

Not all passengers were happy with the privately run ferry system and as early as 1910 public meetings were held to discuss the possibility of a municipal ferry. As the years passed the complaints of unreliability, long waits to cross the water with cars, and excessive rates appeared in letters to the editor of the San Pedro *News-Pilot.*

There was also discussion from time to time about the advisability of constructing a bridge to span the channel. Over the years the water between Wilmington and Terminal Island and Long Beach and Terminal Island had been successfully crossed with draw-bridges, pontoon bridges and bascule bridges. It was suggested as early as 1909 that although traffic was light at the time, the developing harbor would someday be a world port and a bridge from San Pedro to Terminal Island would be a good idea.

The 1939-40 *Annual Report* of the Port of Los Angeles told of a study made by the Harbor Department with the view to improving the present service by the construction of modern ferry terminals at both ends of the channel crossing. It also provided for adequate service through the operation of a modern, efficient ferryboat capable of maintaining fast service. The construction of a new terminal at Sixth Street on property acquired from the Southern Pacific Railroad was considered. This site would shorten the existing diagonal ferry route.

A notice inviting bids for furnishing ferry service between San Pedro and Terminal Island was issued in September 1940, by the Los Angeles Harbor Department. The service was to provide for two double ended ferryboats, a double deck main ferry which would operate for twelve hours on an eight to ten minute round trip schedule and a single deck auxiliary ferry operating the other twelve hours on a less frequent schedule. The Harbor Department was to construct ferry terminals at the foot of Sixth Street in San Pedro and at Terminal Way on Terminal Island. These were to replace the two existing landings in San Pedro at Fifth Street and First Street. The new route would be shorter and more direct.

It wasn't until September 2, 1941, that a municipal ferry system began to operate between Sixth Street in San Pedro and Fourth Street on Terminal Island. The fondly remembered passenger-automobile ferry Islander, traveling at six knots (about seven miles per hour), crossed the one thousand feet of channel water during the busy daylight hours and the wooden foot-ferry, the Ace, carried shipyard and cannery workers back and forth during the nighttime shifts. They were to do this for twenty-two years.

Development of the canneries had brought hundreds of Japanese fishermen and workers to the island where some of them had built their homes. At the outbreak of World War II all civilian residents were removed from the island to make way for the expansion of the shipyards and navy installations. Added to these facilities were the U.S. Coast Guard Station, the Custom House, Immigration Station and Federal Prison, all bringing people to "somewhere" and greatly increasing the flow of traffic to the island. In the decades after World War II the further development of the Port of Los Angeles as a world trade center created the need for faster and more convenient access to the now greatly enlarged (by fill) Terminal Island.

As early as 1937 Senator (later Governor) Cuthbert Olsen had introduced a bill to the California State Senate which provided for connecting the highways of Long Beach and Los Angeles Harbor with a tunnel under the channel. Although the bill was not successful it was an indication that there was great concern for an efficient flow of traffic in the harbor area.

Site Selection

Before the bridge there were plans for the tunnel or tube. For 36 years, beginning in 1922, all surveys and legislation dealt with the proposal of connecting San Pedro to Terminal Island under the channel rather than over. In 1923, as evidenced by a map, there were six different proposed connections. One began at Sixth and Palos Verdes Streets and travelled much the same path as the ferry, crossed the channel and terminated at Terminal Island at Berth 236. The northernmost suggested juncture dated 1922-1935 was situated just north of the present bridge. On the San Pedro side it was reached by Regan Street, went across the turning basin, met with Terminal Island just south of Ferry Street and terminated at Ocean Avenue.   The southernmost proposed route began at Fourteenth and Palos Verdes Streets, made a northerly bend at Beacon Street, continued at a diagonal across the channel, connected with Terminal Island at Berth 238, made a curve and terminated at South Seaside Avenue and Cannery Street.

Vincent Thomas' working copy of the map which shows the proposed tube and bridge locations.

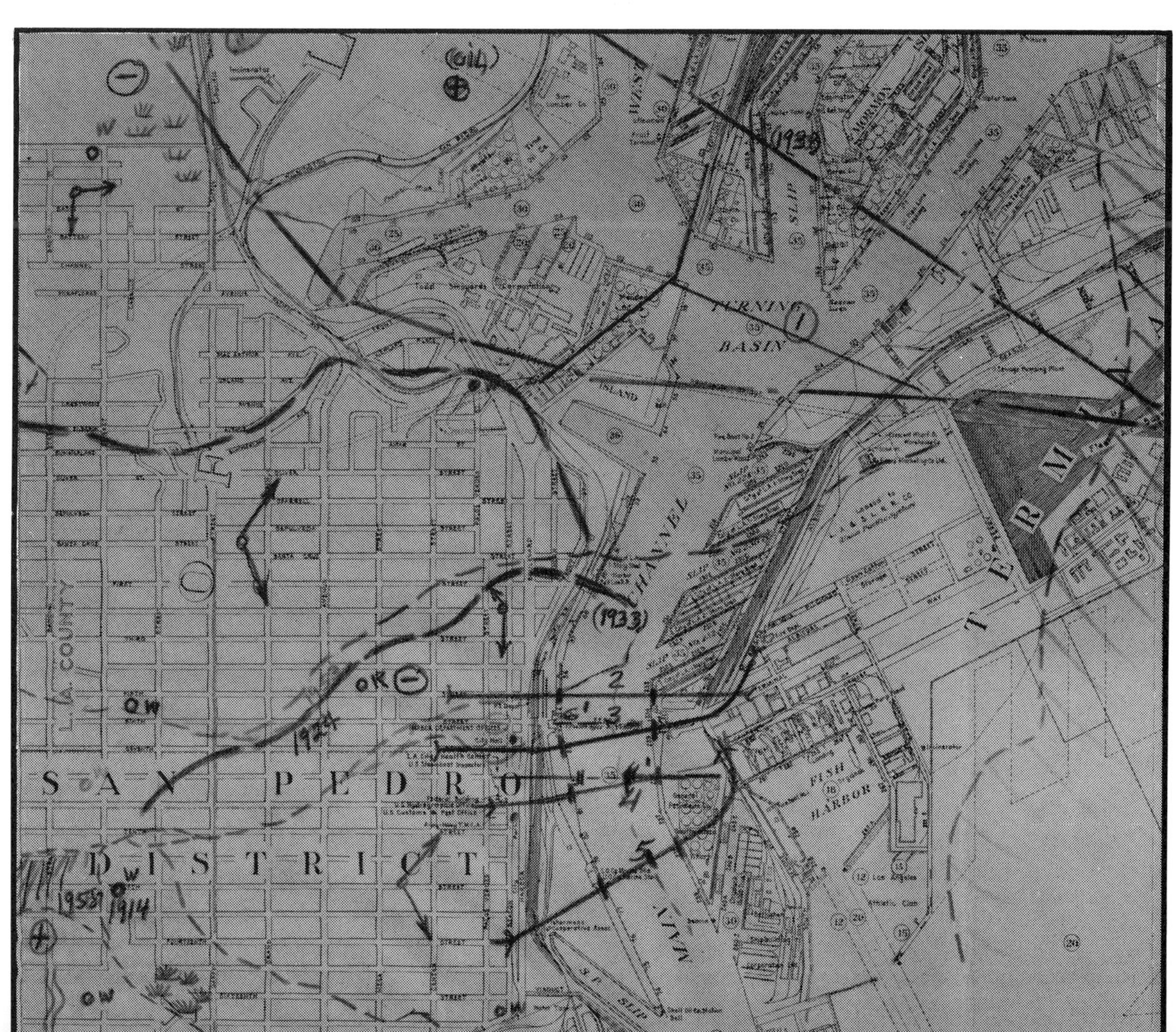

On June 12, 1957, Governor Goodwin J. Knight signed AB 1152, which enabled building the San Pedro-Terminal Island tube. William Neal and Lewis Arnold, Assistant City Attorneys for Los Angeles, and Assemblyman Vincent Thomas witnessed the event.

The first piling being driven at Seaside and Ferry Streets on Terminal Island. The Vincent Thomas Bridge was the first ever to have its anchorages built entirely on pilings.

Legislation

Legislation for the tube was first introduced in 1937 by then Senator, later Governor, Cuthbert Olson, who authored Senate Bill 590 that proposed a "T" tunnel to connect San Pedro and Terminal Island. This statute most likely referred to the northernmost tube proposal on the map as a "T" tunnel.

HR 46, authored by Assemblyman Vincent Thomas in 1948, was the first tunnel measure to pass both houses. The resolution asked that the Division of Highways investigate the "necessity and feasability of constructing a San Pedro, Wilmington, Terminal Island, and Long Beach tunnel to connect the Harbor and Long Beach freeways."

This resolution was followed by AB 25 in 1949, calling for the tunnel route to come under the state highway system; HR 32 in 1952, asked the Toll Bridge Authority to study a means of faster transportation between San Pedro and Terminal Island; AB 2721 in 1953, provided funds for a complete survey; AB 657 in 1955, established a method of financing, constructing, operating, maintaining and insuring the tube; AB 1905 in 1957, defined the toll tube and AB 1152 in 1957, enabled the means for financing the project.

With these numerous pieces of legislation laying the groundwork for completion of a tunnel connecting San Pedro and Terminal Island, the engineers approached Thomas with the proposal that a four lane bridge could be constructed for just a little more than the two lane tube. Consequently, legislation had to be rewritten to accommodate building a bridge instead.

In 1958, Thomas authored AB 58 proposing "a bridge, with at least four lanes, from San Pedro at or near Boschke Slough to Terminal Island." Although the bill passed, the bridge could not begin until surveys were made to determine not only the actual number of cars using routes to and from Long Beach but, also, an estimation and evaluation of the potential traffic through and future growth of the community. It would also take into consideration the future growth of the industries, residences and commercial establishments in the area.

The Board of Harbor Commissioners passed a resolution in 1958 to support the bridge project and agreed to furnish rights-of-way to any land under its jurisdiction free of charge. These rights-of-way valued at $2.5 million were contributed by the Los Angeles Harbor Department.

The bill to authorize construction and funding of the bridge originally went to the Assembly and Senate with $2 million each to come from the City and County of Los Angeles' share of gas tax revenue, $5 million from the southern counties gas tax revenue and $15 million in revenue bonds. The bill was amended in the Senate by Senator John Murphy to write out the revenue coming from the twelve southern counties. As amended the bill passed the Senate and Assembly and was signed by Governor Goodwin J. Knight on April 17, 1958.

On April 30, 1958, the California Toll Bridge Authority considered authorization of construction of the bridge. The Bridge Authority needed to approve the sale of revenue bonds to finance the project before the bridge's construction could begin. Tolls would be charged to cover the principal and interest on the bonds.

In 1959, Thomas authored two pieces of legislation to finance the bridge. AB 2359 was unsuccessful but AB 2356 was passed and provided the funding.

In April 1959, Lieutenant Governor Glenn M. Anderson won approval by the California Toll Bridge Authority for a resolution calling for the State Highway Commission to loan $4.5 million of gas tax funds to complete the financing of the bridge. The loan would be in addition to $5 million from the State, $9 million in revenue bonds, $1 million from the Los Angeles Harbor Department, $1 million from Los Angeles City and $2 million from Los Angeles County.

State engineers presented the results of their study on the proposed bridge to the community of San Pedro on June 25, 1959, in the San Pedro High School Auditorium. The purpose of the meeting was to record local support or opposition to the project. With legislation, financing and community support prepared, groundbreaking for the bridge took place on May 28, 1960. The shovels were lifted by Governor Edmund G. Brown and Assemblyman Vincent Thomas with State Senator Richard Richards and Lieutenant Governor Glenn Anderson looking on.

Although ground was broken, the actual construction of the bridge was delayed because no bids were received for the $7 million in revenue bonds. Bids were initially opened in September 1960, but by October 3rd, no bids had been received by the State Toll Bridge Authority, who immediately readvertised the bonds for sale.

By November 1960, there were still no buyers for the bonds. The $14 million in construction bids submitted in September 1960, by Guy F. Atkinson Company of Long

Beach to build the substructure for $2,576,465 and Yuba Consolidated Industries Incorporated and Kaiser Steel Corporation of San Francisco to build the superstructure for $11,482,000, had to be rejected. The second offer for bids for the revenue bonds called in early October had been rescinded when the state decided to look into making the bonds more attractive to buyers.

On April 26, 1961, at the Bank of America headquarters in San Francisco, the bonds were delivered by Assistant Attorney General E. J. Funke, acting for the California Toll Bridge Authority, to William Ireland of First Western Bank representing Allen & Company of New York, head of the underwriting syndicate that bought the bonds. The Bank of America acted as the Bridge Authority's fiscal agent in paying interest and returning bonds from toll revenues. Construction of the bridge could finally begin.

A crawler crane is standing by ready to hoist another steel section being prepared by the riggers for the San Pedro side tower. Prefabricated pieces are in the foreground.

Aerial view of the west side of the channel cable bent and on-ramp supports graduating down to shorter heights to the point where the bridge will join the Harbor Freeway. Front Street is seen in the background as it curves around Knoll Hill. Todd Shipyard is at top right.

Construction

Work on the substructure of the Vincent Thomas Bridge began in May 1961, by Guy F. Atkinson Company the original bidder for the job. The firm contracted to build the substructure and all roadwork on both sides of the channel, the toll plaza and the administration building.

The cable anchorages for the bridge were set on pilings making it the first ever to be constructed entirely on pilings. The anchor blocks, weighing 23,000 tons, set on 228 prestressed piles, were driven diagonally 50 feet into the ground in the direction of the resultant force to support the anchorage and in addition create an upward and backward thrust to prevent the block from sliding forward. The bootings were designed to hold the cable weight of approximately 10,000 tons.

The first steel erected for the bridge, a piece of 24-ton fabricated steel, was placed on January 30, 1962. In attendance to witness the work were Assemblyman Vincent Thomas; Councilman John S. Gibson; Jack Carlson, Kaiser Steel; State Resident Bridge Engineer, J.M. Curran; and Robert O. Valentine, Yuba Erectors. The contract for the superstructure work went to the original bidder, Kaiser Steel Corporation, which sublet the tower erection to Yuba Erectors, and the cable work to John R. Roebling's Sons.

All told, three such structures were set on the foundation and bolted together to form the bottom of the tower. Each leg contains seven sections with each tower weighing 1,480 tons; 990 steel piles, supporting 145 tons each, were driven to support the towers. The bridge collected another "first" by being the first welded suspension bridge in the United States as no rivets were used. Both the towers had been fabricated, completely assembled, dismantled for shipment at Kaiser Steel's Montebello plant and then erected at the bridge site to insure perfect fit.

Once completed the towers stood 35 stories above the water and contained 6.2 million pounds of reinforced steel and 22 million pounds of structured steel. In addition, the towers sway outward in a 360-degree arc following the sun across the sky. The engineers designed the towers with this movement to allow the bridge to expand and contract evenly.

With the towers in place, work was ready to begin on the cable spinning, one of the most intricate jobs on the bridge. Two one-inch cables were carried across the channel and lifted into place. On these additional cables were placed and eventually the catwalks

Workman atop the mainland tower adjusting the cable that will support the spinner's catwalk. The transit sheds in mid-rear and right are on the site of the former American President Lines. The berth, built after World War II, featured a clock tower which is still in operation today.

At left is the cable anchor block with the bent, which holds the cable at the proper angle to support the span, and towers in view behind. Also the spinner's catwalk is visible between the tower uprights.

This early 1960s photo shows the bridge construction in progress as well as the surrounding areas in either static or developmental stages. At the base of the San Pedro tower, middle foreground, is where the S.S. Princess Louise is berthed today. The newly dredged inlet to the immediate left and the land to the right under the span were the former site of the Kerckhoff-Cuzner Mill and Lumber Company. To the left of the Terminal Island tower are the fireboat station and the Berth 228 finger pier, both of which have been demolished recently. The Hammond Lumber yard was once located just right of the tower.

Looking from San Pedro, the cables are complete with wires hanging down awaiting attachment of the deck that can be seen on the ground between the bent and the tower.

Last portion of the decking over the water being raised from a barge below. The safety nets are hanging under the span, to prevent workmen from falling once paving has begun.

A painter begins the final step of the project. Painting and repainting the bridge has proved to be a perpetual task.

were hung from them. Spinning, which refers to the "spinning-wheel-like movement" of the carrier pulleys, was the next step in construction. The cable was checked for slackness during the spinning operation with final adjustments taking place at night to obtain a more even temperature over the entire span.

The cables contain 4,028 lengths of 3/16-inch galvanized wire, that form a 13 5/8-inch main cable. In addition the main cables consist of 19 strands each made up of 212 wires. They were designed to withstand a 90 mile-per-hour wind which as a safety factor is double the 45 mile-per-hour necessary for a bridge of that kind.

The deck of the bridge was the final major project in the construction. Steel reinforcing bars were fastened in a tight framework under which plywood was placed to hold the concrete while it cured. Once the concrete was set the plywood was removed.

Paving the bridge is a little more difficult than paving a highway since the deck must be balanced. To insure proper balance a chart was designed that showed the alternating sections to be paved. At one end a section would be paved and next a corresponding section at the other end was paved. The pavement consisted of two layers: a base of lightweight concrete, then the top layer of special grout. Altogether 9,000 cubic yards of the light material that weighs 30% less than regular concrete and made of expanded volcanic aggregate were used. As the paving was laid the weight gradually drew the span down into its present arc under which an eighteen story tall ship could pass under at high tide with five feet to spare. Finally, the cables were wrapped and the entire span painted.

The Bridge Department of the California Division of Highways designed the structure. In charge of the design team were Senior Engineer William J. Jurkovich; R.C. Cassano, approaches; D.M. Fraleigh, towers; E.T. Latham, anchorage and cable bent; R.L. Whitaker and W.H. Howard, suspension. Resident engineers on the project were Senior Engineer J.M. Curran, who also worked on the Oakland Bridge; J.O. MacNeill, assistant; J.B. Poppe, assistant for cable spinning; R.P. Bartley, assistant for steel erection; M. Valentine, assistant for surveys and triangulation; and H. Wolfe, assistant for concrete. Thirty-five state engineers were present during the construction to inspect every inch of the work.

The bridge is 6,060 feet long. The main suspended span is 1,500 feet long and its side spans are 500 feet long. The towers are 365 feet tall. The bridge consists of 92,000 tons of Portland cement, 13,000 tons of lightweight concrete, 14,100 tons of steel and 1,270 tons of suspension cable.

With the bridge in left background, Assemblyman Vincent Thomas addresses the crowd during the dedication ceremonies held on Saturday, September 28, 1963, during the Fisherman's Fiesta. The masts of two decorated fishing boats can be seen just behind Mr. Thomas.

The toll station on Terminal Island is ready for opening.

Dedication

The dedication of the Vincent Thomas Bridge took place during the weekend of the 13th annual Fisherman's Fiesta on September 27 to 29, 1963. A Bridge Executive Committee, made up of forty state and community leaders, planned the ceremonies. The weekend was promoted as "the biggest weekend in the history of the Los Angeles Harbor," and Vincent Thomas was named the Honorary Skipper for the Fiesta.

Although the Fiesta opened the festivities for the weekend on Friday night, the dedication began on Saturday morning, September 28th, at 9:30 a.m. At that time a one and a half hour long parade began at Fifteenth Street and Harbor Boulevard, proceeding down Harbor Boulevard. It passed a reviewing stand of visiting dignitaries at Seventh Street and continued down Harbor to O'Farrell Street where it turned into the Consolidated Terminal lot. About seventy marching units consisting of military and youth bands, color guards and drill teams participated in the parade with Governor Edmund G. Brown serving as Grand Marshal.

The parade was the prelude to the formal ceremonies that took place at 11:30 a.m. that day in the Consolidated Terminal lot. Dedication speakers included Thomas, former Governor Goodwin J. Knight, Secretary of State Frank Jordan, Comptroller Alan Cranston, State Senator Thomas Rees, Mayor Samuel W. Yorty, Representative Augustus Hawkins and Supervisor Burton Chace. A squadron of Navy Phantom II jets flew from San Diego and swooped low over the bridge.

After the ceremonies the visiting officials attended luncheons on board boats in the harbor. At 6 p.m. a reception and banquet were held in the newly completed $14 million Consolidated Marine Terminal next to the bridge. In attendance were close to 1,600 people.

Opening

The actual opening to traffic of the Vincent Thomas Bridge took place Friday, November 15, 1963, at midnight. Festivities preceding the caravan across the bridge included a banquet and the last ferry ride of the Islander. Later newspaper stories have stated that President Kennedy had been invited to attend these ceremonies but missed the opportunity, because he was preparing for a trip to Dallas. However, the opening of the bridge was almost purely a "Pedro" affair.

The evening began on Thursday, November 14th, at 7:30 p.m. with a dinner at the Yugoslav-American Club attended by more than 300 people. Walter B. MacArthur, who was in charge of the ferry operation, entertained the audience with anecdotes about the ferry. John Brunac played the accordion during the dinner often inspiring the guests to sing along. Chamber of Commerce President Al Atchison spoke about the bridge as the first of many large projects planned for San Pedro.

The last ride of the Islander took place at 9:30 p.m. that evening. More than 300 people were on the ferry as it went down the Cerritos Channel and into the Long Beach inner harbor and back again. A few teary-eyed passengers disembarked including some who clutched souvenirs such as the wooden lid of a fire bucket from the vessel.

The ribbon-cutting ceremony at the Vincent Thomas Bridge took place at 11:45 p.m. Just before the ribbon was to be cut a car roared up and over the bridge and then returned. At first some feared that someone was trying to beat the first crossing but it turned out the man was in charge of the installation of the bridge's lights and was just making a final inspection. The Chamber of Commerce Secretary Ken Jordan handed a pair of wooden shears to Vincent Thomas who cut the ribbon as Al Atchison, Harbor Commissioner George Watson and Senior Bridge Engineer Jack Curran looked on. The group then got into their cars and were escorted across the bridge by California Highway Patrol motorcycle officers. The caravan of 100 cars was led by a bus carrying 40 persons including the Assemblyman and members of his family. When the caravan arrived at the toll plaza R. L. Hathaway, who was in charge of the bridge operation, was waiting. He accepted the toll for the bus (for both that and the return trip) from Thomas as Thomas' two children, Vincent Jr., eight, and Mary Virginia, twelve, and their cousin, Alan Burin, stood by. All drivers in the caravan received toll scrip as part of the $10 per-person cost of the crossing.

The auto ferry Islander making its last run at 9:30 p.m., November 14, 1963.

Vincent Thomas, at right, prepares to cut the ribbon to signify the opening of the bridge. To his immediate left is Realtor Al Atchison. Other celebrants look on.

Vincent Thomas strikes this happy pose while anticipating the completion of the bridge.

On September 28, 1963, the Los Angeles City Council presented the Honorable Vincent Thomas with this scroll and the first toll scrip book in commemoration of the dedication of the San Pedro-Terminal Island bridge.

Toll Collection and Maintenance

When the bridge first opened a toll of 25 cents was taken each way, but starting in 1983, a charge of 50 cents was taken for west bound traffic only and east bound traffic crossed free. This change has accounted for a loss in revenue of no more than $25,000; however, fewer toll takers are needed thereby reducing labor costs.

Currently a debt of $8.4 million remains to be paid on the loans made from the state highway fund and the County and City of Los Angeles. The bond issued in 1963 and an additional bond issued in 1968 to connect the bridge to the freeway have been retired.

The bridge will no doubt always be a toll bridge because the California Transportation Commission has the authority to continue collecting fees to pay for the operation, maintenance and insurance after all bonds and loans from bridge construction have been discharged.

Over the years changes have been made to the bridge. In 1987, numerous projects were undertaken including reconstruction of the toll plaza in May. The work included widening the roadway, installing a traffic signal and removing two toll booths from the east bound (free) direction. New fencing was installed on the bridge in the latter part of the year. A twelve foot chain link fence was put up on the main span and a six foot one installed along the approach ramps to prevent debris from being tossed over the bridge and hitting workers below and also to deter suicides.

Painting the bridge is a routine and continuous maintenance job. Once completed it is time to begin again. It requires 1,500 gallons of zinc, 500 gallons of primer and 1,000 gallons of green paint to cover the span.

Future plans for the bridge include installation of automated toll taking machines and resurfacing of the deck.

This 1960s photo shows many buildings which are no longer standing. Looking toward East San Pedro and Long Beach one sees finger piers accommodating two break bulk freighters. With the impact of containerization these piers have vanished along with the ferry building upper center. The barren area, upper left, known as Reeves Field, is now covered with imported autos.

Mid 1960s photo focuses on imported cars in temporary storage awaiting movement to auto outlets. This area has since been vacated to allow room for the ever-expanding container business. Most of the buildings and tanks surrounding the vehicles have been demolished.

Impact on the Harbor

Although the transition from the horse and buggy days into the 20th century jet age did involve a tumultuous relationship between the ILWU and PMA, especially during the decade of the sixties, most if not all became a profitable reality. The irony of this historic change is that contrary to the conventional wisdom both the planning for the Vincent Thomas Bridge and the introduction of the machine were being legislated and formulated in the mid 1950s. The "bridge to nowhere" really had a destination. It only remained for the forces and resources to be set in motion.

In its 1960 annual report to member companies, the Pacific Maritime Association, which represents American and foreign shipowners, stevedore companies, terminal operators and related maritime enterprises, included a variety of nationwide comments concerning the historic Memorandum of Understanding. The P. M. A., as it is commonly called, in the previous year had negotiated with the International Longshoremen's and Warehousemen's Union an unprecedented agreement which would have far-reaching impact on all longshore work.

Said the San Francisco Chronicle:

> American industry and labor have achieved something new and highly desirable in the West Coast longshore agreement for a mechanization fund. In this contract the Pacific Coast shipping industry, which has been one of the nation's most backward technologically, approaches the problem of mechanization in a highly intelligent way that yields benefits to everybody. Of few labor contracts can so much be said. West Coast ports will be made more competitive by mechanization. Cargoes will move faster. Trading volume should increase. The back-breaking work of the stevedore is lessened, without economic penalties to the worker. A new era arrives on the waterfront.

Tourism and through traffic aside, there is no doubt that the Vincent Thomas Bridge, mechanization and modernization have forever changed the Los Angeles Harbor. The evolution will continue as the process of innovation is introduced and obsolescence occurs.

Terminal Island, or East San Pedro (a term rarely used anymore), has witnessed an almost complete topographical upheaval. Passing on into oblivion have been the finger piers built during the 1920s, Hammond Lumber yard, Vegetable Oil Products, National Metals, Union Pacific Railroad Depot, Ferry Building, Fire Station (replaced), Fire

A spectacular aerial photo of the Los Angeles Harbor taken in 1963 shows vast empty areas later developed for a variety of cargo handling uses. The bridge and its approaches are in various stages of construction.

Boat Station, Terminal Island Elementary School, Reeves Field, small boat yards and assorted buildings and warehouses. All of these have vanished within the last thirty years. The early 1900s saw the gradual disappearance of the Brighton Beach summer homes of the wealthy and World War II hastened the demise of the Japanese fishing village. It has been the voracious growth of containerization which primarily has replaced a major part of former Terminal Island enterprise and services. Imported autos and bulk cargoes are also part of the cause and effect. A yet to be perfected idea is the on dock rail loading system for all shipping projects.

The aging Cerritos Channel Badger Avenue double-leaf bascule bridge connecting Wilmington to Terminal Island, erected in the 1920s, sits in a drawn position. It has served the auto, truck and rail system well through the years, and its historic importance is well documented. A new rail access is needed for this crossing area.

In 1959, 145,000 containers and 1,044,785 autos were handled by Pacific Coast ports. Last year 75,658,551 containers and 19,209,803 autos crossed the bull rails. The ports of Long Beach and Los Angeles combined were credited with 39,134,901 containers or 51.7% of the entire west coast. The automobiles shared between the two ports amounted to 44.8% or 8,600,969.

There are some depressed situations on Terminal Island. The fishing industry is in a state of doubtful recovery and the shipbuilding and repair business is shaky on both sides of the channel. Only time will tell their ultimate fate.

Whatever foolish notions some people may have had about the Vincent Thomas Bridge can be laid to rest. On the 25th Anniversary of the bridge's opening we come to realize that the impossible dream has become one of the most vital links in the growth of the Los Angeles and Long Beach Harbors.

The President Pierce of American President Lines is berthed at San Pedro 93 with a hammerhead crane handling containers. The transit shed adjacent to the ship has since been razed and replaced with a new, modern passenger facility.

Former California Governor Ronald Reagan signs the bill which enabled completion of the Harbor Freeway-bridge link as Vincent Thomas, left, and an unidentified observer, right, look on.

Harbor Freeway Completion

On March 16, 1968, with about 300 persons in attendance, Governor Ronald Reagan led the groundbreaking ceremonies as keynote speaker for the final link between the Harbor Freeway and the Vincent Thomas Bridge. Construction had actually started one month and a half before but conflicting time schedules delayed an earlier celebration. Attending were numerous dignitaries including Senator Ralph C. Dills, Assemblyman Vincent Thomas, Assemblyman L.E. Townsend and city and county officials. Music was provided by the Long Beach Municipal Band. Site of the ceremony was at Elberon Street and Grand Avenue, the approximate location where Route 47 and Highway 11 connected. Mr. Thomas was quoted as saying "The freeway connection culminates a long continuous effort to complete the freeway system from downtown Los Angeles to the Harbor Area."

At that time bridge crossing of trucks and autos was already averaging 11,500 cars per day. The last vital coupling of an essential roadway was in place.

The Harbor Freeway project was one of three approaches to the bridge. A $5,451,000 contract was awarded to Guy F. Atkinson Company of Long Beach. A half mile stretch was started at Battery Street and curved southeast to meet the bridge near Harbor Boulevard. The third phase was a simultaneous one mile extension of Harbor Freeway from Battery Street to O'Farrell Street. This included a 2.5 mile segment of widening the freeway from four to six lanes between Pacific Coast Highway and Battery Street. The official completion date according to Caltrans headquarters was September 4, 1970. It had been scheduled to finish in spring of the same year.

To this day San Pedrans are still disgruntled that the Harbor Freeway was not named the San Pedro Freeway. After all, the San Diego, Santa Ana and Santa Monica Freeways, to name a few, cross county and city lines. Either way, the Harbor Freeway still ends in San Pedro, the home and district of former Assemblyman Vincent Thomas who through his leadership helped make possible the proliferation of a modern maritime industry.

An early 1960s view of a dredger's clam about to release its dripping load onto a barge. In the background a freighter under foreign flag has converted its weather deck for handling containers.

A 1954 aerial view of the Cerritos Channel shows the drawn Badger Avenue Bridge. Directly behind this span is the Henry Ford Bridge.

The Bridge and the Community

The Vincent Thomas Bridge has played an important part in the modernization of the Los Angeles Harbor. The purpose was to improve communication between the mainland and Terminal Island and, in that regard, transport of vehicles and people has increased greatly and at the same time has been efficiently and quickly handled. The long term financing of the bridge and its freeway link is being met with the receipts of tolls.

Modernization of the waterfront and its businesses over the past several decades has incorporated innovations which save time, labor and money. The great bridge is the most conspicuous of the innovations while containerization of cargo is probably the most pervasive. Rebuilding and deep-dredging of the waterfront and the channels have afforded increased accessability and adaptability while buildings and institutions have been brought up to an efficiency level comparable with that of other leading ports of the world.

The actual ferry trip across the channel took only several minutes but there usually was a boarding time delay for cars. The actual turnaround for the boat was 15 minutes, but a commuter might have to wait longer if the ferry filled up before one's turn to drive aboard. Today, the trip to Terminal Island is still only several minutes but there is no waiting. If no bridge or tube had been built, with the increased traffic of recent years, a veritable fleet of ferryboats would have to be rushing back and forth across the channel, the passage hampered by the many ships and other craft passing up and down.

The bridge has also provided easy access between the port, the City of Long Beach and other communities to the east and down the coast. During the ferryboat days before 1963, commuters to Long Beach were likely to travel via Alameda, Anaheim and B Streets. This was usually a quick trip by pre-freeway standards. For non-drivers the journey was a half-hour by Pacific Electric street car. The use of the bridge has shortened auto and bus travel time to 10 or 15 minutes.

Substitution of the bridge for the downtown San Pedro ferry had the apparent effect of bypassing the town, but this was only partly true. The completion of the Harbor Freeway was already bringing a flow of visitors to Ports O' Call tourist complex, beaches, parks and harbor. This traffic increased greatly over the years with the expansion of Ports O' Call, the addition of new museums and the Cabrillo Marina.

The bridge has served Terminal Island commuters as an efficient access to the Federal installations of the Naval Base, Coast Guard Station, Federal Prison, Immigration and Customs House; canneries and boats of the fishing industry; light industry; and dock and harbor facilities. It has accommodated the development of cargo shipping and the auto import transit system.

Struggling, picturesque but deteriorating, the old downtown district, fanning west from Sixth Street and Harbor Boulevard, was affected by removal of cross-channel traffic at the foot of Sixth Street. Business declined more sharply after 1963 and in eight years the district was razed. The past seventeen years has seen a gradual revival of business due to urban renewal and the Beacon Street Redevelopment Project.

Harbor and waterfront business has found the convenience of the bridge not only functionally valuable but an effective image-builder. The shimmering, majestic, green structure is a spectacular backdrop for the booming cruise ship industry. Vincent Thomas' bridge, without a doubt, has had much to do with the positive feeling every visitor has of this fascinating world port, as it provides a picture of unsurpassed beauty from both land and sea.

Vincent Thomas

Visko Tomas and son John (I) are seated and wife Vinca is standing with their daughters Zorka and Jane with young Vince at his father's knee.

Vincent Thomas

Vincent Thomas' parents immigrated to the United States from Yugoslavia. His father, Visko Tomas, was born in Seget in 1873 and his mother Vinca, born in 1881, was a native of Brac. Vince, their third child, was born in Biloxi, Mississippi on April 16, 1908. The Tomases had eight children.

Vince's father, a blacksmith, specialized and was renowned for the fancy New Orleans style grillwork he created. Before World War I his father travelled to Oakland, California. There he found a better opportunity working for the Southern Pacific Railroad as a "round". When established he wrote asking his wife to sell everything in Pass Christian, Mississippi and to join him in Oakland. This the family did.

During the Spanish influenza epidemic in 1918, the family was visiting in San Pedro and Mr. Tomas contracted the illness. Upon diagnosis all flu patients were immediately transferred to Cleveland Mortuary, which had been converted into an emergency hospital. This facility was located at Ninth Street and Pacific Avenue where Trax now does business. A short time after being admitted Visko died.

Vinca moved her family to San Pedro so that they could be closer to relatives. She died in 1925 of diabetes. Mr. and Mrs. Tomas and their son Nick are buried in Oakland. Two other sons died, one by drowning and the other from diptheria, and are interred in Pass Christian.

Vince, his sisters Lena and Jane with brother John II moved in with his sister Zorka and her husband, Tom Nizetich, a fisherman. Thomas attended Fifth Street School and while registering at the grammar school, a teacher asked his name. He replied,"Tomas", but she wrote Thomas. The name stuck and his brother and sisters also adopted the surname. While at San Pedro High School, Coach Carl Haney noted Vince's athletic ability and encouraged his participation in football, basketball, baseball and boxing.

Conscientious and concerned, Vince sold newspapers to help support his brothers and sisters. While in high school he made and sold "snake oil", a rubdown compound. This smelly and burning concoction of boiled king snake fat, oil, peppermint and wintergreen sold well to the local athletes. It was from this venture that Vince acquired his nickname "Snake Oil" Thomas. He also augmented his income by boxing at $2.50 a fight.

Graduating as an Ephebian in 1928, Vince earned an athletic scholarship to Santa Clara University. While there he coached boxing and other sports. In 1932, he acquired a Bachelor's Degree in Philosophy. Vince attended Loyola University from 1934 through 1936, and though he earned his Jurisprudence degree, he did not practice law.

After graduating from Santa Clara, Thomas worked for French Sardine, now known as Star-Kist Foods, Incorporated. He also lobbied in Sacramento for the fishing industry and various other causes. In 1940, Assemblyman Reeves died. Forrest McDaniels, a local haberdasher, and officials at French encouraged Vince, then 32, to run for public office.

Because Thomas always felt that Los Angeles treated San Pedro like a "stepchild" he campaigned championing "Secede from L.A.". Sentiment was with him, he was victorious and became the new assemblyman.

During his tenure Vince preferred to serve as a committee member because he felt he could be more effective concentrating on his district's problems. Thomas was appointed to the Human Resources and Labor Relations Committees and the Pacific Marine Fisheries Commission.

Vince served as vice-chairman of the Fish and Game Committee for ten years. In 1955-56, he was the Minority Floor Leader. The first chairmanship Thomas accepted was in 1963 to the important Legislature's watchdog over State spending, the Joint Legislative Audit Committee. Later he chaired the Intergovernmental Relations Committee.

While in the Assembly Vince represented the 68th District well. He worked to improve the lot of the California commercial fishing industry, and to provide for recreation, sports, the aged, the blind, the handicapped, transportation and education. Thomas introduced the legislation which provided that the Harbor Freeway terminate in San Pedro and for the bridge accesses, as well as the original enactment authorizing the creation of the California State College (now University) of Dominguez Hills.

The Assemblyman is considered "boxing's saint" as he dedicated himself to upgrading the sport by writing and introducing legislation regulating it, as well as providing pension benefits for ex-fighters. His office was filled with mementos of his boxing and legislative careers. He regularly clipped newspaper stories on both and kept them in huge scrapbooks.

The greatest accomplishment of his career was the completion of the bridge that crosses the Main Channel of the Los Angeles Harbor from San Pedro to Terminal Island. To honor his foresight and untiring work that culminated in its construction, the Legislature enthusiastically passed the Concurrent Resolution 131 naming the bridge for Vincent Thomas. This bill was authored by Assemblyman Augustus Hawkins and Senator Hugh Burns.

Thomas was a presidential elector in 1940 and 1944 and also a delegate to the National Democratic Party Conventions in 1948 and 1960. Vince instituted the first ombudsman program in the state to better serve his constituency.

As the Assemblyman with the longest consecutive service in 1963, Vince became known as the "Dean of the Assembly". He still holds the record as he served the 68th District for a total of 38 consecutive years or nineteen terms. During 1969, Thomas suffered a heart attack and his recovery took six weeks.

A bitter defeat at the polls in 1978 ended a long and admirable career. Vince had hoped to serve a total of 40 years and then retire. The loss brought on an earlier retirement.

Vincent and Mary Thomas renewing their wedding vows at their 25th anniversary mass with Frank and Anita Trani in attendance. Reverend Monsignors George M. Scott and George M. Gallagher officiated.

The Thomases' daughter, Mary Virginia Thomas Haughey, her husband Richard with sons Michael and Nicholas.

The Thomases' daughter-in-law and son, Cindy and Vincent Thomas, Jr.

His Family

Mary DiCarlo was dating a local dentist when Anita and Frank Trani, owners of Trani's Shoe Store, were planning a dinner party. Needing to "even up the table," Anita called Mary inviting her to attend, stressing that it was not to be a matchup. Her dinner partner would be Assemblyman Vincent Thomas, who would be returning from a Chicago meeting. Mary didn't feel she should go, but after some convincing agreed.

That night after an evening of discussing politics, Vince asked if he could take Mary home. He did so and as he was about to leave, remarked that he would like to show her and the Tranis his cellar and wine collection. He promised to be in touch soon.

Time passed and Mary didn't hear from Vince. Then one evening she went to dinner with her family at the Palms Restaurant in Wilmington. Thomas was there with several friends. As a courtesy Pietro DiCarlo, her father, sent drinks over to his table. Vince stopped by their table as he was leaving. He told Mary that he hadn't forgotten about their getting together with the Tranis to see his wines, and that she would hear from him soon.

In September, about two weeks later Vince called to ask Mary to attend a football game at the Coliseum. The dentist was working that day, so she agreed to go. They then began to see more of each other.

One evening when Vince and some other legislators were visiting the DiCarlo family Senator Randolph Collier, Dean of the Senate and "father of the California highway system" asked Mr. DiCarlo, "Would it be okay if Vince and Mary could get married?" Mary says that her parents froze momentarily and that finally her father said, "I guess so." Vince never proposed personally.

Within six weeks of their first date, on November 5, 1947, Vince and Mary eloped to Ensenada, Mexico and were married there as they wished to avoid a large "political" wedding. They had two children, Mary Virginia (Ginny) and Vincent, Jr. Both met their spouses while attending Santa Clara University in Northern California. Ginny married Richard Haughey of Santa Clara in 1973, and they have two sons, Michael, 12, and Nicholas, 8. In 1978, Vince Jr. married Cindy Trobitz from Eureka, California.

Throughout their life together the only competition Mary sensed was the proposed tube or the bridge. Vince always believed that San Pedro needed mass transportation, particularly a fast means to cross the Main Channel. He met constantly with legislators, community representatives, businessmen and officials from Kaiser Steel to fulfill these dreams.

Vincent Thomas was a very private and hospitable individual, who worked diligently for his district. He did not seek laurels. On January 31, 1980, he died of a massive heart attack, leaving what could be considered his memorial – the Vincent Thomas Bridge.

Bibliography

(California Division of Highways). *The Vincent Thomas Bridge,* (1963).

Caltrans."Financial Summary of the Vincent Thomas Bridge," August 8, 1988.

Hager, Anna Marie. "Terminal Island's Glamorous Port," *Shoreline,* San Pedro Bay Historical Society, February 1983.

Houston, John. "Early San Pedro Ferry Service," *Shoreline,* San Pedro Bay Historical Society, January 1979.

___________.*The San Pedro City Dream*, Parts I, II, III.

Los Angeles City commendations.

Los Angeles Harbor Department. *Notice Inviting Bids,* "Ferry Service - Los Angeles, Specification Number 1000."

___________. *Annual Reports* 1913, 1939-1940, 1946, 1947.

___________. *Worldport Los Angeles - Annual Reports 1987.*

Los Angeles *Times,* 1958-1988.

Pacific Maritime Association .*Annual Report* 1960, 1987.

Press-Journal and *Harbor Mail,* October 3, 1962.

Reid, Patricia. "This Modern Green Giant Celebrates Its Silver Anniversary," Manuscript, (1988).

Sacramento *Bee,* March 6, 1977.

San Pedro Bay Historical Society Archives.

San Pedro *News-Pilot,* 1958-1987.

Thomas, Mary. Oral interviews in San Pedro July 14, 21, 22, 28, 1988.

Vincent Thomas Dedication Committee. "The Vincent Thomas Bridge fact sheet," (1963).

Walsh, Donald A. *Los Angeles Harbor,* Manuscript.

Silver Anniversary Celebration

To commemorate the silver anniversary of the opening, the Vincent Thomas Bridge 25th Anniversary Committee chaired by Labor Arbitrator Marston Chavez was formed. The group was composed of volunteer representatives of San Pedro's business and residential communities.

Caltrans planned the closure of the bridge to vehicular traffic from 12:30 to 6 p.m. on November 6, 1988, to accomodate the following festivities: a brunch at the Princess Pavillion; a short program featuring national, state and local dignitaries; an old car parade; bands; an afternoon of pedestrian walkovers; helicopter and vintage airplane flyovers; and a fireboat and yacht parade. The day concluded with individuals lining the bridge holding lights that illuminate the span. The proceeds raised before and during the day's activities will finance the permanent lighting of the Vincent Thomas Bridge.